NUCLEAR PHYSICS, NEW ATOMIC BOMB, BIONIC ARM

Written and created by

ANDREW IGLA

Copyright © 2017 by Andrew Igla.

ISBN:	Softcover	978-1-5434-4547-3
	eBook	978-1-5434-4548-0

All rights reserved. No part of this book may be reproduced or transmitted in any form or by any means, electronic or mechanical, including photocopying, recording, or by any information storage and retrieval system, without permission in writing from the copyright owner.

Any people depicted in stock imagery provided by Thinkstock are models, and such images are being used for illustrative purposes only.
Certain stock imagery © Thinkstock.

Print information available on the last page.

Rev. date: 11/14/2018

To order additional copies of this book, contact:
Xlibris
1-888-795-4274
www.Xlibris.com
Orders@Xlibris.com
766525

Nuclear Physics, New Atomic Bomb, the Bionic Arm

CHAPTER 1

Atomic Physics, Atomic Bomb

"DEDICATED TO HELEN LIM A JEWISH 23 YEAR OLD BRITISH AGENT FRIEND WHO WAS MURDERD SO I ANDREW IGLA COULD FINISH THIS IDEA THE NEW ATOMIC BOMB WITH VARYING INPUT DISTANCE OF A NEUTRON MODERATOR ENTERING THE DETENATION CHAMBER ,KNOWN TO BRIGADER MARCUS SMITH OF THE AUSTRALIAN MILITARY."

May she rest in peace. July 2017.

Nuclear Physics, New Atomic Bomb, the Bionic Arm

Let us look at simple mathematics of nuclear reactors first.

Infinite Multiplication Factor, k_∞ ....
Four Factor Formula ....
Fast Fission Factor, (ε) ....
Resonance Escape Probability, (p) ....
Thermal Utilization Factor, (f) ....
Reproduction Factor, (η) ....
Effective Multiplication Factor ....
Fast Non-Leakage Probability ($\mathcal{L}_f$) ....
Thermal Non-Leakage Probability ($\mathcal{L}_t$) .....
Six Factor Formula ....
Neutron Life Cycle of a Fast Reactor .....

Nuclear Physics, New Atomic Bomb, the Bionic Arm

In a **nuclear decay reaction** an unstable nucleus emits radiation and is transformed into the nucleus of one or more other elements. The resulting daughter nuclei have a lower mass and are lower in energy (more stable) than the parent nucleus that decayed.

Binding energy is equal to the **decrease in potential nuclear energy** of the nucleons when they come together. This is equivalent to the work done on the nucleons by the strong nuclear force.

$$\Delta m = Zm_p + (A - Z) m_n - m_{nucleus}$$

where:

Δm = mass difference
m_p = mass of a proton
m_n = mass of a neutron
$m_{nucleus}$ = mass of the formed nucleus
Z = proton number or atomic number
A = nucleon number or mass number

Nuclear Physics, New Atomic Bomb, the Bionic Arm

Binding energy is the above $\Delta m\, c^2$

Which is the lower energy of the combined mass minus the individual energy of mass components times -1.

The below chemical reaction consumes a neutron and slows down **or delays a nuclear chain reaction.** This will be shown in a controlled nuclear engineering science system to delay normal uranium fission or to put it simply **delay mathematically for the first time in science the detonation of an atomic bomb.** This is a wanted system where detonation is delayed.

"Uranium 238"

$$^{238}_{92}U + n \rightarrow\, ^{239}_{92}U + \gamma \xrightarrow{\beta^-}\, ^{239}_{93}Np \xrightarrow{\beta^-}\, ^{239}_{94}Pu$$

Nuclear Physics, New Atomic Bomb, the Bionic Arm

Let us look at normal fission or uranium decay and create mathematics to control this new detonation.

"Uranium 235"

Fission

In the fission reaction the incident neutron enters the heavy target nucleus, forming a compound nucleus that is excited to such a high energy level ($E_{exc} > E_{crit}$) that the nucleus "splits" (fissions) into two large fragments plus some neutrons. An example of a typical fission reaction is shown below.

$$^{1}_{0}n + ^{235}_{92}U \rightarrow \left(^{236}_{92}U\right)^{*} \rightarrow ^{140}_{55}Cs + ^{93}_{37}Rb + 3\left(^{1}_{0}n\right)$$

So here three 3 neutrons are created to create a nuclear chain reaction.

Mathematics is needed to control this type of nuclear detonation via a delayed nuclear chain reaction.

Let us go on a journey now here.

Nuclear Physics, New Atomic Bomb, the Bionic Arm

Start with…

General theory about neutrons breaking down to protons and electrons and anti neutrinos noted here first.

neutron ===> proton + electron + a neutrino

See previous statements

"Uranium 235"

"Uranium 238"

Therefore thermal neutrons are more likely to cause uranium-235 to fission than to be captured by uranium-238. If at least one neutron from the U-235 fission strikes another nucleus and causes it to fission, then the chain reaction will continue. If the reaction will sustain itself, it is said to be critical, and the mass of U-235

required to produce the critical condition is said to be a critical mass.

Nuclear Physics, New Atomic Bomb, the Bionic Arm

Fast fission is fission that occurs when a heavy atom like uranium 238 absorbs a high-energy neutron, called a fast neutron, and splits.

Most nuclear fuels contain heavy fissile elements that are capable of nuclear fission, such as **uranium-235** or **plutonium-239**.

Plutoneium reaction.

$$^{239}_{93}Np \rightarrow \,^{239}_{94}Pu + \,^{0}_{-1}\beta + \,^{0}_{0}\bar{v}$$

Beta decay is the emission of electrons of nuclear rather than orbital origin. These particles are electrons that have been expelled by excited nuclei and may have a charge of either sign.

Certainly neutron absorbtion will be worked out in differential mathematical calculus for this project not mathematical algebra like nuclear reactor theory.

Nuclear Physics, New Atomic Bomb, the Bionic Arm

In nuclear engineering, a **neutron moderator** is a medium that reduces the speed of fast neutrons, thereby turning them into thermal neutrons, in temperature equilibrium with its surroundings, capable of sustaining a nuclear chain reaction involving uranium-235 or a similar fissile nuclide.

The release of neutrons from the nucleus requires exceeding the binding energy of the neutron.

Moderation is the process of the reduction of the initial high kinetic energy of the free neutron. Since energy is conserved, this reduction of the neutron kinetic energy takes place by transfer of energy to a material known as a moderator. It is also known as *neutron slowing down*, since along with the reduction of energy comes a reduction in speed.

This repartition of the neutrons in a pressurised water reactor shows the different roles played by slow and fast neutrons. Slow neutrons are responsible for most of nuclear fission with uranium 235 and therefore help sustain the chain reactions. Fast neutrons, on the other hand, play a small role in fission but can transform nuclei of uranium 238 into fissile plutonium 239. Other neutrons are lost when they are captured by 'sterile' nuclei or when they escape from the reactor as they slow down. In one of the most remarkable phenomena in nature, a **slow neutron** can be captured by a uranium-235 nucleus, rendering it unstable toward **nuclear fission**. A fast **neutron** will not be captured, so **neutrons** must be **slowed** down by moderation to increase their capture probability in **fission** reactors.

Nuclear Physics, New Atomic Bomb, the Bionic Arm

The fissioning of an atom of uranium-235 in the reactor of a nuclear power plant produces two to three neutrons, and these neutrons can be absorbed by uranium-238 to produce plutonium-239 and other isotopes. Plutonium-239 can also absorb neutrons and fission along with the uranium-235 in a reactor.

This will be used in the detonation time delay in this project.

So there are thermal neutrons, fast neutrons, creation of neutrons, and absorption of neutrons.

Reactor theory use generation to explain neutron generation.

In this project we will use generation of neutrons and a new system of mathematical calculus formulae using **time, space, and temperature, mass of various uranium types.**

Nuclear Physics, New Atomic Bomb, the Bionic Arm

The atom density of uranium-235 is 4.83×10^{21} atoms/cm3. The atom density of uranium-238 is 4.35×10^{22} atoms/cm3.

These figures will be used in space mass calculations.

The average number of neutrons produced per fission (v) in this project will be converted to per time in detonation.

Mass of uranium introduced to the detonation system is controlled.

A **neutron poison** (also called a **neutron absorber** or a **nuclear poison**) is a substance with a large neutron absorption cross-section, in applications such as nuclear reactors.

In this detonation project the space is divided up into volumes of neutron absorption and neutron creation not focusing on a cross-section area. In both these volumes the temperature of the space is involved in the calculus equation of delayed detonation.

In this project neutrons per total volume times second will be used.

In this project neutrons per total area times second will be used.

Temperature per total volume times second will be used.

Temperature per total area times second will be used.

Neutron creation probability density or neutron absorption probability density per total meter times second will be used in this project.

The probability density mathematics will be a travelling wave in space and time.

Neutron creation probability density or neutron absorption probability density per total meter squared times second will be used in this project.

Nuclear Physics, New Atomic Bomb, the Bionic Arm

Conversion Factors:
1 amu = 1.6606 x 10 -27 kg
1 newton = 1 kg-m/sec2
1 joule = 1 newton-meter
1 MeV = 1.6022 x 10-13 joules

Back ground points begin…
Creation of alpha particle

$$^{234}_{92}U \rightarrow {}^{230}_{90}Th + {}^{4}_{2}\alpha + \gamma + KE$$

Beta decay is the emission of electrons of nuclear rather than orbital origin.

$$^{239}_{93}Np \rightarrow {}^{239}_{94}Pu + {}^{0}_{-1}\beta + {}^{0}_{0}\bar{\nu}$$

Back ground points end.

Nuclear Physics, New Atomic Bomb, the Bionic Arm

Types of neutron collision outcomes

U-235 (n,γ) U-236: Capture of a neutron and ejection of γ.

U-235 (n,f) Fission Products: Neutron fission

Fe-56 (n,n) Fe-56: Scattering of a neutron.

First type collision…

There can be elastic scattering where the neutron slows down and the uranium speeds up but no nuclear fission happens here.

Second type collision…

There can be inelastic scattering where the uranium is excited in energy state and has some kinetic energy and the neutron has kinetic energy reduced during the collision.

Nuclear Physics, New Atomic Bomb, the Bionic Arm

The excited uranium here eventually does not go through nuclear fission but emits harmless gama rays and becomes less exited.

Third type collision…

The neutron collides with the uranium and gets absorbed by the uranium and becomes U-236 from U-235 uranium. The uranium decays in energy state via gamma rays.

Fourth type collision…

The neutron collides with the uranium and gets absorbed and the result is nuclear fission not gamma rays **because of a yet undiscovered type of different excitation of the uranium than collision type three.**

Nuclear Physics, New Atomic Bomb, the Bionic Arm

Fission is about a type of excitation of the uranium 235 not fully understood yet as why would it not just produce gamma rays like collision three 3.

The approach Andrew Igla is using is to design calculus mathematics to model nuclear detonation in fission and control this unknown concept as calculus mathematics is used for controlling nuclear submarines of USA heating and cooling in the nuclear reactor, that I have contributed too already in my life in nuclear physics in Washington D.C. USA . Calculus mathematics is allowing bypass of knowledge of nuclear physics to control a nuclear system.

The time delay of nuclear detonation can only be understood and controlled by calculus mathematics.

Nuclear Physics, New Atomic Bomb, the Bionic Arm

Fission produces fast moving neutrons which cannot be used in further fission until the neutrons are slowed down from fast moving neutrons to thermal neutrons.

In nuclear reactor theory typically in Washington D.C. in modern times certain physics modelling and mathematics algebra is used. Okay we are not using this physics model of neutrons.

All we are interested in is fast moving neutrons released from fission and slow moving neutrons to be used by fission.

Also neutron absorption not relating to fission, uranium 238 converting to plutonium, will be mathematically modelled by my design of delayed nuclear detonation known from here on in this document as **the nuclear delay system.**

Mathematical variables used

$r = f(x,y,z)$. t = time. X axis, y axis, z axis

Uranium 235 density u235(r,t)

$U235(r,t) = U1\{\cos(br-wt) + j\sin(br-wt)\}$

a travelling wave in space and time of system volume [detonation volume] of the nuclear delay system is the probability density of finding a uranium particle 235.

Neutron n1 density n1(r,t)

$N1(r,t) = n1\{\cos(kr-wt) + j\sin(kr-wt)\}$

Uranium 238 density u238(r,t)

$U238(r,t) = U2\{\cos(br-wt) + j\sin(br-wt)\}$

b radians per meter is the same for all uranium types.

Nuclear Physics, New Atomic Bomb, the Bionic Arm

Uranium in the nuclear detonation volume is a travelling wave and unlike particles or atoms per meter is proportional to the probability density of an uranium atom being in space at a time t.

Gamma, alpha, beta radiation is part of

U235(r,t) U238(r,t) n1(r,t) formulae as a wave component

The formulaes including gamma, alpha, beta radiation in the detonation chamber become

$N1(r,t) = n1\{ \cos(kr-wt) + j\sin(kr-wt) + \cos(kr - w1t) + j\sin(br-w1t)\}$

This is real Einstein modelling of detonation. To explain.

Nuclear Physics, New Atomic Bomb, the Bionic Arm

The formulas including gamma, alpha, and beta radiation in the detonation chamber become

$$U238(r,t) = U2\{\cos(br-wt) + j\sin(br-wt) + \cos(kr-w1t) + j\sin(br-w1t)\}$$

U235 stays the same as already mentioned as fission equation where no large amount of gamma, alpha, beta radiation is occurring.

The mathematics is a surgical bypass of physical chemistry of gamma, alpha, beta radiation concerning only detonation physics.

W1 is the radian <u>constant frequency</u> of gamma, alpha, beta radiation above =

$2\pi f1.w = 2\pi f.$

Let us look at

$\partial U235(r,t)/\partial t \quad \partial U235(r,t)/\partial r$

$\partial U235(r,t)/r^2/\partial t \quad \partial^2 U235(r,t)/r^2/\partial r^2$

Terms.

Let us look at

$\partial U238(r,t)/\partial t \quad \partial U238(r,t)/\partial r$

$\partial U238(r,t)/r^2/\partial t \quad \partial^2 U238(r,t)/r^2/\partial r^2$

Terms.

Let us look at

$\partial n1(r,t)/\partial t \quad \partial n1(r,t)/\partial r$

$\partial n1(r,t)/r^2/\partial t \quad \partial^2 n1(r,t)/r^2/\partial r^2$

Terms.

New physics stored at the pentagon Washington D.C. checked out by the pentagon as correct new physics.

The formulae are...

$\int U235 dr$ times $\int U235 dt$ times $\{dr/dt\}$

is the arrival rate of atoms of uranium at r,t.

"This is new in nuclear physics and is the Igla nuclear arrival rate. 2017 July."

Signature digital: Avraham Menachem mendel.

Nuclear Physics, New Atomic Bomb, the Bionic Arm

This new physics is derived from arrival rates of a particle from probability density in Schrodinger's equation. 1927 from physics max born 1927 an Austrian Jewish person.

Igla1 equation.

$\int U235 dr$ times $\int U235 dt$ times $\{dr/dt\}^2$ times n; n = mass of uranium 235, = E = number of neutrons in uranium 235 times $(3/2) K_B T$

This is the expected temperature T at a point in the detonation chamber at r,t.

K_B is boltzman's constant.

When T exceeds a maximum temperature $T^O C$ MAX the bulk head of the atomic bomb tears open breaking the graphite cooler of neutrons releasing radioactive radiation from the uranium fuel.

Nuclear Physics, New Atomic Bomb, the Bionic Arm

The dr/dt formulae is equal to = w/b

Reminding

$U235(r,t) = U1\{\cos(br-wt) + j\sin(br-wt)\}$

Implying from igla1 equation...

$n(1/w)(1/b)e^{2\{br-wt\}} (w/b)^2$ = Energy =E = number of neutrons in uranium 235 times (3/2) $K_B T°c$; after integrating the igla1 equation.

Since $\int U235 dr$ **times** $\int U235 dt$ **times** $\{dr/dt\}$

is equal to the number uranium atoms per second at point 'r' at time 't' known from Schrodinger's equation of probability density function and particle arrival rates, the above equation is true physics and correct and Andrew Igla's new physics as the pentagon of Washington D.C. has stated.

Nuclear Physics, New Atomic Bomb, the Bionic Arm

Two things happen in the detonation chamber of my new atomic device heating and cooling like the nuclear reactor of a USA nuclear submarine.

The graphite nuclear neutron cooler slowing neutrons down from nuclear fission plus the heat sinks cooling down the temperature of the new atomic device in this delayed atomic detonation device is what WE are here for.

IMPLYING d^2T/dr^2 a spatial acceleration of temperature in the detonation chamber. Now in nuclear submarines the cooling temperature and heating temperature is kept track of. In my igla device the temperature of the heat sinks is all that is needed to keep track of mechanical bulk head failure as a safety factor is built in.

$dT°c/dt$ is also used in this model, a speed of temperature in time at the heat sink.

Nuclear Physics, New Atomic Bomb, the Bionic Arm

Reviewing our physical chemistry.

Fission

In the fission reaction the incident neutron enters the heavy target nucleus, forming a compound nucleus that is excited to such a high energy level ($E_{exc} > E_{crit}$) that the nucleus "splits" (fissions) into two large fragments plus some neutrons. An example of a typical fission reaction is shown below.

$$^{1}_{0}n + ^{235}_{92}U \rightarrow \left(^{236}_{92}U\right)^* \rightarrow ^{140}_{55}Cs + ^{93}_{37}Rb + 3\left(^{1}_{0}n\right)$$

$$^{238}_{92}U + n \rightarrow ^{239}_{92}U + \gamma \xrightarrow{\beta^-} ^{239}_{93}Np \xrightarrow{\beta^-} ^{239}_{94}Pu$$

Delay of detonation is buy pulling away the dynamic graphite device reducing this graphite device volume slowing down the creation of thermal slower neutrons. This delays and slows down fission.

Graphite volume is gv. Implying…

$\partial gv / \partial t$

Uranium 238 increase slows down fission by absorbing neutrons and releasing gamma rays. Fission creates fast neutrons and increases temperature.. Implying…

$\partial U238(r,t)/ \partial t$ or $\partial U238(r,t)/ \partial r + \partial gv/\partial t + \partial gv/\partial r =$

$A \, \partial^2 T \, °c/dr^2 + B \, \partial T \, °c /dt$

But this is a small equation of heating and cooling.

The heat sinks are fixed in volume as we are interested in detonation in the end unlike nuclear reactors in nuclear submarines but with a mathematical calculated time of detonation.

Instead of varying the heat sink in size I vary the graphite neutron slow down volume creating calculatable number of thermal neutrons.

$\partial n1(r,t)/\partial t, \quad \partial n1(r,t)/\partial r$

Relates to fast neutrons after fission

$A \, \partial U235(r,t)/\partial t + B \, \partial U235(r,t)/\partial r =$

$J \, \partial n1(r,t)/\partial t + \quad K \, \partial n1(r,t)/\partial r$

is a fast neutron creation equation.

Nuclear Physics, New Atomic Bomb, the Bionic Arm

Uranium fission $A\, \partial U235(r,t)/\partial t +$

$B\, \partial U235(r,t)/\partial r\, -$ minus

$\{\, J\, \partial n1(r,t)/\partial t + K\, \partial n1(r,t)/\partial r\,\}$ fast neutron creation $=$

$\partial gv/\partial t\, \{\, A1\, \partial^2 T\,°c/dr^2 + B1\, \partial T\,°c/dt\,\}$ graphite volume rate times temperature change.

Let this be the Igla2 equation.

$A\, \partial U235(r,t)/\partial t +$

$B\, \partial U235(r,t)/\partial r\, -$

$\{\, J\, \partial n1(r,t)/\partial t + K\, \partial n1(r,t)/\partial r\,\}\, =$

$\partial gv/\partial t\, \{\, A1\, \partial^2 T\,°c/dr^2 + B1\, \partial T\,°c/dt\,\}\,.$

Nuclear Physics, New Atomic Bomb, the Bionic Arm

Implying igla3 equation as follows…

A ∂ U235(r,t)/ ∂ t +

B ∂ U235(r,t)/ ∂r −

{ J ∂ n1(r,t)/ ∂ t + K ∂ n1(r,t)/ ∂r } =

∂ gv /∂ t { A1 ∂^2T °c/dr^2 + B1 ∂ T °c /dt } .

^{238}U is a fissionable isotope, but is not fissile isotope. ^{238}U is not capable of undergoing fission reaction after absorbing thermal neutron, on the other hand ^{238}U can be fissioned by fast neutron with energy higher than >1MeV. ^{238}U does not meet also alternative requirement to fissile materials. ^{238}U is not capable of sustaining a nuclear fission chain reaction, because too many of neutrons produced by fission of ^{238}U have lower energies than original neutron.

Implying igla4 equation…

$A \, \partial U235(r,t)/\partial t +$

$B \, \partial U235(r,t)/\partial r -$

$\{ \, J \, \partial [\, n1(r,t) - U238(r,t) \,]/\partial t +$

$K \, \partial [\, n1(r,t) - U238(r,t) \,]/\partial r \, \} \; =$

$\partial gv /\partial t \, \{ \, A1 \, \partial^2 T\,^\circ c/dr^2 + B1 \, \partial T\,^\circ c /dt \, \}.$

Because slown thermal neutrons are not involved with uranium U238 AND fast neutrons never get a chance to become slow neutrons, thermal neutrons, for fission in uranium 235.

But $\partial U235(r,t)/\partial t = \partial U235(r,t)/\partial r$ times $dr/dt = \partial U235(r,t)/\partial r \, (w/b)$

Since $U235(r,t) = U1\{\cos(br-wt) + j\sin(br-wt)\}$

$dr/dt = w/b$; $br-wt = c$; since $b(dr/dt) - wdt/dt = 0$

Igla4 equation can be rewritten ...

$A \, \partial U235(r,t)/\partial r \, (w/b) +$

$B \, \partial U235(r,t)/\partial r -$

$\{ \, J \, \partial [n1(r,t) - U238(r,t)]/\partial r \, (w/b) +$

$K \, \partial [n1(r,t) - U238(r,t)]/\partial r \, \} \, =$

$\partial gv /\partial t \, \{ A1 \, \partial^2 T \, °c/dr^2 + B1 \, \partial T \, °c /dr \, (w/b) \}$

Igla4 equation can be rewritten...

$\partial U235(r,t)/\partial r \, \{ A (w/b) + B \} + [$

$\partial [n1(r,t) - U238(r,t)]/\partial r \, \{ J (w/b) + K \}]$

$= \partial gv /\partial t \, \{ A1 \, \partial^2 T \, °c/dr^2 + B1 \, \partial T \, °c /dr \, (w/b) \}$ called the igla5 equation for nuclear detonation.

Nuclear Physics, New Atomic Bomb, the Bionic Arm

But there are two more physics issues needed here.

1/Delay of detonation.

2/ energy output of detonation.

$\partial gv /\partial t$ is not converted to an 'r' variable equation as there is no dr/dt. There is no increasing and decreasing the graphite unit in space because we are not trying to control temperature like a nuclear submarine in the bulkhead.

There is a steady time delay of detonation wanted here.

Let $\partial gv /\partial t = ka(1 - e^{-k1t})$; This is physics and more thermal neutrons will be created for nuclear fission here as 't' approaches a larger number or time approaches a number Tmax; as the graphite volume increases.

But Boltzmans constant is involved in this physics.

So if we take ...

$\partial\, U235(r,t)/\partial\, r\, \{ A\, (w/b) + B \} + [$

$\quad \partial\, [\, n1(r,t) - U238(r,t)\,]/\partial\, r\, \{ J\, (w/b) + K \}\,]$

$\partial\, gv/\partial\, t\, \{ A1\, \partial^2 T\, °c/dr^2 + B1\, \partial\, T\, °c\, /dr$

as the left hand side of an equal sign and

$gv = a\cos w_0 t$.

Then the right hand side of the equal side

$= d\, (K_B T^O w_0)/d\, r\, (1/r^2)$

If we take gv the introduced space in time of the neutron moderator, slowing down fast neutrons to slow thermal neutrons for nuclear fission with uranium 235, with the formulae

$A(1 - e^{-t\cos(wt-kr)})$ then the following applies…

$\partial \, U235(r,t)/\partial r \, \{ A \, (w/b) + B \} + [$

$\quad \partial \, [\, n1(r,t) - U238(r,t) \,]/\partial r \, \{ J \, (w/b) + K \} \,]$

$\partial \, gv \, /\partial t \, \{ A1 \, \partial^2 T \, °c/dr^2 + B1 \, \partial T \, °c \, /dr$

$= \partial \, U235(r,t)/\partial r \, \{ A \, (w/b) + B \} + [$

$\quad \partial \, [\, n1(r,t) - U238(r,t) \,]/\partial r \, \{ J \, (w/b) + K \} \,]$

$\partial \, gv \, /\partial t \, \{ A1 \, \partial^2 T \, °c/dr^2 + B1 \, \partial T \, °c \, /dr \} =$

$\partial \, U235(r,t)/\partial r \, \{ A \, (w/b) + B \} + [$

$\quad \partial \, [\, n1(r,t) - U238(r,t) \,]/\partial r \, \{ J \, (w/b) + K \} \,]$

$\partial \, gv \, /\partial r \, (dr/dt) \, \{ A1 \, \partial^2 T \, °c/dr^2 +$

$\quad B1 \, \partial T \, °c \, /dr \}$

Implying…

$\partial U235(r,t)/\partial r \{ A(w/b) + B \} + [$

$\quad \partial [n1(r,t) - U238(r,t)]/\partial r \{ J(w/b) + K \}]$

$\partial A(1- e^{-t\cos(wt-kr)})/\partial[t\cos(wt-kr)]$

$\partial[t\cos(wt-kr)]\partial r$ times

$\quad \{ A1\ \partial^2 T\ ^oc/dr^2 + B1\ \partial T\ ^oc/dr \}$

$= d(K_B T^O w)/dr (1/r^2).$

w is the travelling wave frequency.

K_B is boltzman's constant above.

Nuclear Physics, New Atomic Bomb, the Bionic Arm

More matching in physics units is...

U235uranium single particle = u235p...

U235(r,t) = U1{cos(br-wt) + jsin(br-wt)}

w = 2πf...

$\int u235 dr = j(1/b)e^{j(br-wt)}$

$\int u235 dt = -j(1/w) e^{j(br-wt)}$

Kg/u235p ∂[∫u235dr ∫u235dt]/∂t → ∂r [∂r/∂t]³ = $K_B T°C$ w = power surge in physics.

Also note...

r Kg/u235p ∂²[∫u235dr ∫u235dt]/ ∂r² [∂r/∂t]³ = $K_B T°C$ w = power surge in physics.

Nuclear Physics, New Atomic Bomb, the Bionic Arm

Physics units of the right hand side of the equal sign and left hand side of the equal sign match on the previous equations.

Now take temperature of the atomic bomb using Andrew Igla's design of a moving size neutron moderator increasing in size in Andrew Igla's atomic bomb.

There is cooling of the atomic bomb in temperature...

$(\partial r/\partial t) = w/b$. r is the maximum neutron moderator distance minus the neutron moderator distance.

w units radians per second;

b radians per meter ;

$sK_B (\partial T°c/\partial t) (e^{-rcos(br-wt)}) = K_B (\partial T°c/\partial r)(\partial r/\partial t)$ $(e^{-rcos(bx-wt)})$ ... r = distance length of neutron moderator ; now $K_B (\partial T°c/\partial t)(e^{-rcos(bx-wt)}) = K_B (\partial T°c/\partial r)(\partial r/\partial t)(e^{-rcos(br-wt)}) = K_B T°C\, w$

Temperature of atomic bomb heating.

r is the neutron moderator distance.

$$rK_B (\partial T^\circ c/\partial t)(1-e^{-r\cos(br-wt)}) = K_B(\partial T^\circ c/\partial r)$$
$$(\partial r/\partial t)(1-e^{-r\cos(br-wt)}) \ldots r = \text{distance length of neutron moderator.}$$

$$rK_B (\partial T^\circ c/\partial t)(1-e^{-r\cos(br-wt)}) = K_B(\partial T^\circ c/\partial r)$$
$$(\partial r/\partial t)(1-e^{-r\cos(br-wt)}) = K_B T^\circ C\, w$$

representing power surge.

Now in USA NUCLEAR submarines there is a heating up temperature a $T_H^\circ c$, Cooling down temperature $T_c^\circ c$.

BUT in the above equations there is terms…

$$(1-e^{-r\cos(br-wt)}), (e^{-r\cos(br-wt)})$$

But br above = $\vec{b}.\vec{r}$ …in physics… Which is different for heating and cooling of Andrew igla's atomic bomb. Because cooling is not equal to $\vec{T}$ heating allowing the use of one temperature variable mathematically in physics.

Nuclear Physics, New Atomic Bomb, the Bionic Arm

There is an increasing neutron moderator length near the uranium for nuclear reactor heating and a decreasing neutron moderator length near the uranium for nuclear reactor cooling.

Vary in cycle the length of neutron moderator near the uranium allows nuclear reactor cooling and heating in a small space able to exist in a missile fueling the missile with nuclear power with range around the world five time.

If r cooling is at 45 degrees to r heating…

Then the terms…

$$(1-e^{-r\cos(br-wt)})\cdot,\cdot(e^{-r\cos(br-wt)})$$

become…

$$(1-e^{-r\cos(\vec{b}\cdot\cdot\vec{r}\cdot(\cos(\pi/4))-wt)})\cdot,\cdot(e^{-r\cos(\cdot\vec{b}\cdot|\cdot\vec{r}\cdot\cdot-wt)})$$

Nuclear Physics, New Atomic Bomb, the Bionic Arm

Surgically removing the need for defining a sepertate heating and cooling temperature via the space time continuum.

"r heating = r cooling times $\cos(\pi/4)$... from a source point of the neutron moderator."

So let us look at a differential equation of fission with a time space continuum varying neutron moderator length and vector space heating area and different vector space cooling area AND one temperature mathematics variable.

$\int u235 \, dr = j(1/b)e^{j(br-wt)}$

$\int u235 \, dt = -j(1/w) \, e^{j(br-wt)}$

$\partial r/\partial t = w/b$; $\partial t \to \partial r$ means converts.

Nuclear Physics, New Atomic Bomb, the Bionic Arm

The new power Andrew Igla fission equation.

$$[\{Kg/u235p\, \partial[\int u235dr \int u235dt\,]/\partial t \rightarrow \partial r\} - \{Kg/u238p\, \partial[\int u238dr \int u238dt\,]/\partial t \rightarrow \partial r\}]$$

$$[\partial r/\partial t]\,3 \;+$$

$$r\, Kg/u235p\, \partial^2[\int u235dr \int u235dt\,]/\partial r^2\,[\partial r/\partial t]^3 \;+$$

$$K_B\,(\partial T°c/\partial r)\,(\partial r/\partial t)\,(e^{-r\cos(\vec{b}\cdot\vec{r}-wt)})\| \;+$$

note ↗

$$K_B\,(\partial T°c/\partial r)\,(\partial r/\partial t)\,(e^{-r\cos(\vec{b}\cdot\vec{r}\cos(\pi/4)-wt)})$$

$$= K_B T°Cw = \text{powerAndrew Igla fission equation}$$

Uranium fission is a nuclear chain reaction.

Andrew igla represents the nuclear chain reaction in the space time continuum.

U235 uranium particle probability density 235.

Nuclear Physics, New Atomic Bomb, the Bionic Arm

$\partial \int u235 dr \int u235 dt / \partial \int (r^n \{ \vec{i} \cos\pi/n + \vec{j} \sin\pi/n + \vec{k} \tan\pi/n \}) dn$ times [...

$\partial \int (r^n \{ \vec{i} \cos\pi/n + \vec{j} \sin\pi/n + \vec{k} \tan\pi/n \}) dn / \partial r$] is the spatial representation of the nuclear chain reaction in nuclear physics, known as andrew igla space fission equation from here on in $r^n \cos\pi/n$ is the magnitude and direction r(angle) vector.

n = 0 to infinity.

Incorporate this vector approach into the new power Andrew Igla fission equation.

As n changes value a new particle uranium 235 is found going through nuclear fission.

Nuclear Physics, New Atomic Bomb, the Bionic Arm

Nuclear Physics, New Atomic Bomb, the Bionic Arm

CHAPTER 2

The Bionic Arm

Bionic arm with motor nerves used between spine and shoulder and to a smaller degree sensory armpit nerves to the spine.

Goal.

Small magnetic fields from the nerve bundle between the spine and shoulder in **normal motor nerve** movement will be used to control or input electronics to control a titanium arm using d.c. motors.

Batteries will be used to power the bionic arm.

Sensory nerves between the spine and brain may be used to help the person attached to the bionic arm to control a weight or load in the bionic arms hand. A weight in the hand requires mechanical static force and ongoing nerve signals from the spine to the shoulder to hold the weight and load. Using a mechanical

Nuclear Physics, New Atomic Bomb, the Bionic Arm

engineering approach only allows a heavy load to be held up but not a Lighter weight changing in weight like a pitcher of water being poured out changing the weight dynamically. In a mechanical engineering approach only jerking of the bionic arm would occur as the weight held by the hand changes weight and forces the person to look at the weight or load object in his hand as the water pours out of the pitcher.

Motor nerve analysis is done with Fourier mathematics looking at the intensity and frequency of the nerve signals.

Internal medicine, medically, the nerves to the fingers, hand, forearm, upper arm are giving off magnetic signals in the same spatial area between the spine and shoulder.

Nuclear Physics, New Atomic Bomb, the Bionic Arm

Fourier mathematics will separate the upper arm, forearm, hand nerve magnetic signals with a microprocessor in the bionic arm.

Finger motor movement can be done with a mechanical engineering approach only not using the motor nerves between the spine and shoulder. Or not.

Sensory nerves play a part in finger movement but in any case this is only the introduction phase of the document.

The bionic arm microprocessor will have digital memory and have a few thousand gentle movements able sometimes to carry a light weight like an egg or heavy weight like a bowling ball.

Nuclear Physics, New Atomic Bomb, the Bionic Arm

Movement in rotation of the upper arm is complicated with angle movements and strength being involved. This will be a challenge. Angle shoulder movements alone require moving the weight of the bionic arm only.

Angle shoulder movements holding a weight require strength and angle movements.

Angle shoulder movements with a varying weight like a pitcher of water being poured out require eyes focused on the bionic arm or sensory artificial intelligence or both.

Sensory nerves from the spine or other human hand can be used.

Nuclear Physics, New Atomic Bomb, the Bionic Arm

The magneic fields from the brachial plexus nerve bundle group of nerves are digitally stored on the micro computer of the bionic arm.

Fourier mathematics will be used to separate each nerve signal in the goup of nerves, the brachial plexus. This will allow control of the upper arm, forearm, hand, finger.

The brachial plexus is a complex intercommunicating network of nerves formed by spinal nerves **C5, C6, C7, C8 and T1**

- it supplies all **sensory innervation** to the **upper limb** and most of the **axilla**, with the exception of an area of the medial upper arm and axilla, which is supplied by the intercostobrachial nerve T2
- it supplies all **motor innervation** to the muscles of the **upper limb** and **shoulder girdle**, with the exception of trapezius, which is supplied by the spinal accessory nerve IX

- it also supplies **autonomic innervation** to the upper limb by intercommunicating with the stellate ganglion of the **sympathetic trunk** at the level of **T1**, where it gains sympathetic fibres which supply specialist functions:
    - **pilomotor** – stimulates **contraction of arrector pili muscles** within hair follicles, making hairs stand on end
    - **sudomotor** or **secretomotor** – stimulates **production of sweat** from sweat glands

The axillary **nerve** also carries **sensory** information from the **shoulder** joint.

A electric field, electromagnetic field transmitting to the axillary nerve around the shoulder joint will provide sensory nerve signals from the bionic arm. This will allow the bionic arm to rotate at the shoulder and carry a time varying weight starting as heavy as a bowling ball or pitcher of water being poured out.

The **axillary nerve** or the circumflex **nerve** is a **nerve** of the human body that originates from the brachial plexus (upper trunk, posterior division, and posterior cord) at the level of the **axilla** (**armpit**) and carries **nerve** fibers from C5 and C6.

Electronic weighing instruments can manage time varying weight, like a pitcher of water pouring out water, giving electronic signals to a micro computer giving electromagnetic, electric field signals to **axillary nerve near the arm pit.**

End sensory nerves.

Nuclear Physics, New Atomic Bomb, the Bionic Arm

Basic squid equations

The total current can be written then as

$$i = 2I_c \cos\left(\frac{\pi\Phi}{\Phi_o}\right) \sin\left(\varphi_1 + \frac{\pi\Phi}{\Phi_o}\right)$$

$$\Phi = \Phi_{ext} + LI_{cir}$$

The circulating current is given by $I_{cir} = (i_1 - i_2)/2$

The total flux can then be written as

$$\Phi = \Phi_{ext} + \frac{LI_c}{2} \sin\left(\frac{\pi\Phi}{\Phi_o}\right) \cos\left(\varphi_1 + \frac{\pi\Phi}{\Phi_o}\right)$$

sin(a+b)-sin(a-b)=2cos(a)sin(b)

Electrical model of a squid.

External magnetic field time varying like a nerve signal's magnetic field causing voltage time varying output. 'x' s in diagrams are josephson junctions.

There are direct current squids and radio frequency squids.

In a radio frequency squid only one josephson junction is used.

In medicine the r.f. radio frequency squid is used "normally" **In this project** a direct current squid may be used.

V out d.c. = kI_b

graphs dc squid

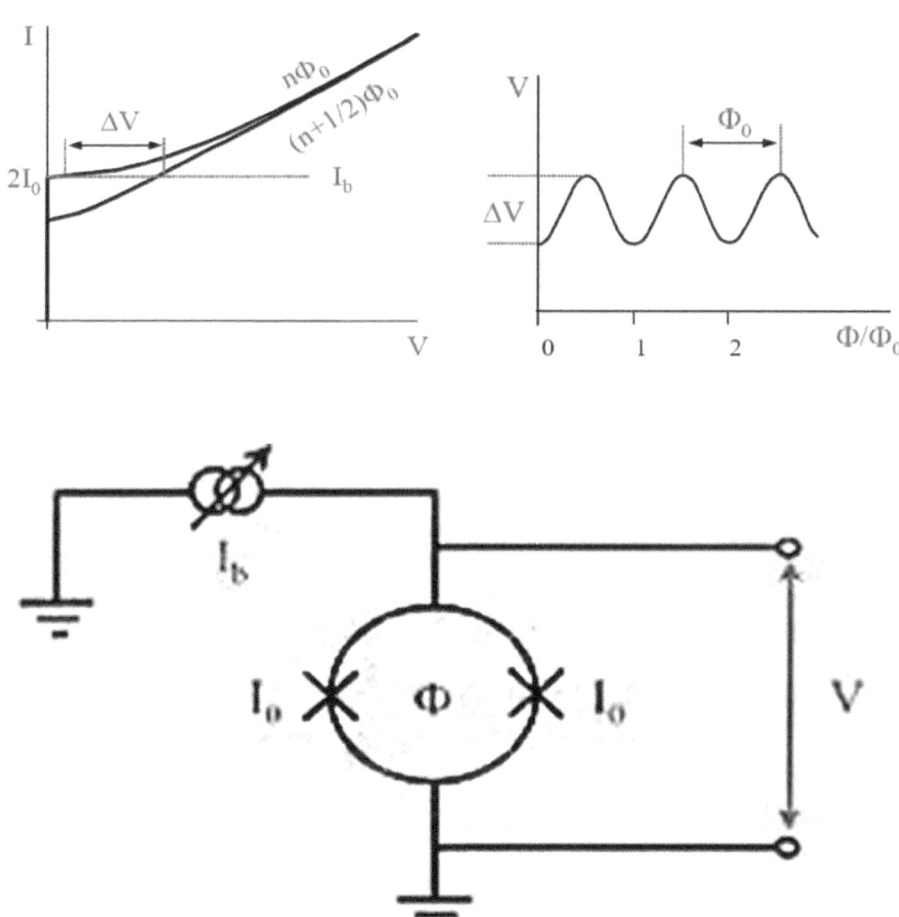

Electrical currents must exceed I_o in each superconducting arm.

SQUIDs are usually made of either a lead alloy (with 10% gold or indium) and/or niobium, often consisting of the tunnel barrier sandwiched between a base electrode of niobium and the top electrode of lead alloy. A radio frequency (RF) SQUID is made up of one Josephson junction, which is mounted on a superconducting ring. An oscillating current is applied to an external circuit, whose voltage changes as an effect of the interaction

The **Josephson effect** is the phenomenon of supercurrent—i.e. a current that flows indefinitely long without any voltage applied—across a device known as a **Josephson junction** (JJ), which consists of two superconductors coupled by a weak link. The weak link can consist of a thin insulating barrier.

Nuclear Physics, New Atomic Bomb, the Bionic Arm

In a Josephson junction, the nonsuperconducting barrier separating the two superconductors must be very thin. If the barrier is an insulator, it has to be on the order of 30 angstroms thick or less. If the barrier is another metal (nonsuperconducting), it can be as much as several microns thick. Until a critical current is reached, a supercurrent can flow across the barrier; electron pairs can tunnel across the barrier without any resistance. But when the critical current is exceeded, another voltage will develop across the junction. That voltage will depend on time--that is, it is an AC voltage. This in turn causes a lowering of the junction's critical current, causing even more normal current to flow--and a larger AC voltage.

The external magnetic flux from a nerve signal is then measured.

Where I_o is the critical current and φ the phase difference between the macroscopic wave functions of the cooper pairs relative to the two superconductors and V is the voltage across the junction

$$I = I_0 \sin \varphi$$

$$= \varphi_1 - \varphi_2$$

$$\varphi_1 - \varphi_2 = 2\pi\frac{\Phi}{\Phi_0} = 2\pi\frac{\Phi_e + LJ}{\Phi_0} \qquad (3)$$

Where φ_1 and φ_2 are the phase differences of the superconducting wave functions across the two junctions. $\Phi = \Phi_e + LJ$ is the total flux threading the SQUID loop given by the external flux Φ_e and the self flux produced by the screening current circulating into the SQUID loop with an inductance L. The circulating current can be expressed as $J = (I_1 - I_2)/2$.

$$\frac{\Phi - \Phi_{ext}}{\Phi_0} = \frac{LJ}{\Phi_0}$$

In a rf squid ib = acoswt and one josephson junction is not there. 'x'

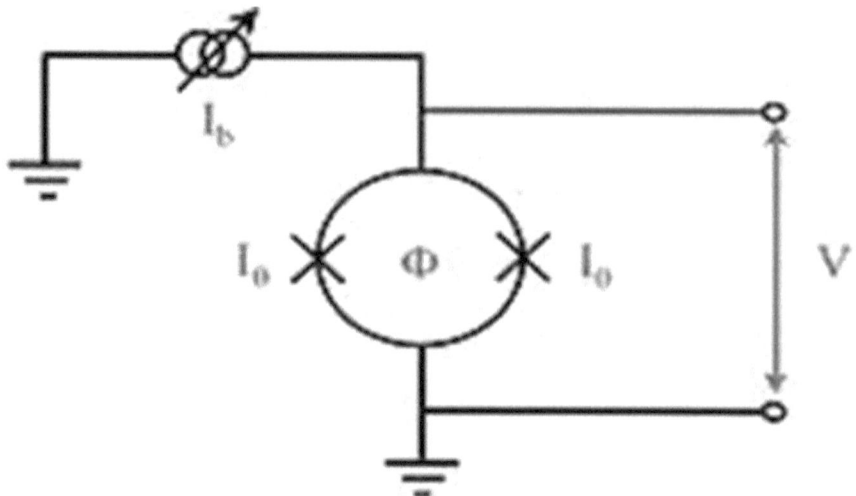

The electric current in the josephson junction is a continous current proportional to the sine of the phase difference between the two superconductors.

The phase variation of a superconductor of a josephson junction can be linked to the external magnetic flux from a nerve signal.

The frequency of this AC voltage is nearly 500 gigahertz (GHz) per millivolt across the junction. So, as long as the current through the junction is less than the critical current, the voltage is zero. As soon as the current exceeds the critical current, the voltage is not zero but oscillates in time. Detecting and measuring the change from one state to the other is at the heart of the many applications for The frequency of this AC voltage is nearly 500 gigahertz (GHz) per millivolt across the junction. So, as long as the current through the junction is less than the critical current, the voltage is zero. As soon as the current exceeds the critical current, the voltage is not zero but oscillates in time. Detecting and measuring the change from one state to the other is at the heart of the many applications for Josephson junctions. Josephson junctions.

The frequency of this AC voltage is nearly 500 gigahertz (GHz) per millivolt acr The frequency of this AC voltage is nearly 500 gigahertz (GHz) per millivolt across the junction. So, as long as the current through the junction is less than the critical current, the voltage is zero. As soon as the current exceeds the critical current, the voltage is not zero but oscillates in time. Detecting and measuring the change from one state to the other is at the heart of the many applications for Josephson junctions. oss the junction. So, as long as the current through the junction is less than the critical current, the voltage is zero. As soon as the current exceeds the critical current, the voltage is not zero but oscillates in time. Detecting and measuring the change from one state to the other is at the heart of the many applications for Josephson junctions.

The frequency of this voltage is nearly 500 gigahertz (GHz) per millivolt across the junction. So, as long as the current through the junction is less than the critical current, the voltage is zero. As soon as the current exceeds the critical current, the voltage is not zero but oscillates in time. Detecting and measuring the change from one state to the other is at the heart of the many applications for Josephson junctions.

An SFQ pulse is produced when magnetic flux through a superconducting loop containing a Josephson junction changes by one flux quantum, Φ_0 as a result of the junction switching. SFQ pulses have a quantized area $\int V(t)dt = \Phi_0 \approx 2.07 \cdot 10^{-15}$ Wb $= 2.07$ mV ps $= 2.07$ mA pH due to magnetic flux quantization, a fundamental property of superconductors.

$\Phi_0 \approx 2.07 \cdot 10^{-15}$ Wb $= 2.07$ mV ps $= 2.07$ mA pH due to magnetic flux quantization

Note for the following page

$$i = 2I_c \cos\left(\frac{\pi\Phi}{\Phi_o}\right) \sin\left(\varphi_1 + \frac{\pi\Phi}{\Phi_o}\right)$$

$$\Phi = \Phi_{ext} + \frac{LI_c}{2} \sin\left(\frac{\pi\Phi}{\Phi_o}\right) \cos\left(\varphi_1 + \frac{\pi\Phi}{\Phi_o}\right)$$

The magnetic field of a nerve is detected as a phase signal of a cosine mathematical function. <u>Fourier</u> is the wrong approach. m(t) = nerve signal.

Cos {f(t) = m(t) + sin(m(t)/j)cos((m(t)/j) + β)|}

ϕ ext = external = total nerve bundle magnetic signal.

Nuclear Physics, New Atomic Bomb, the Bionic Arm

Now suppose the external flux is further increased until it exceeds $\phi_0/2$, half the magnetic flux quantum. Since the flux enclosed by the superconducting loop must be an integer number of flux quanta, instead of screening the flux the SQUID now energetically prefers to increase it to ϕ_0. The screening current now flows in the opposite direction. Thus the screening current changes direction every time the flux increases by half integer multiples of $\phi_0/2$. Thus the critical current oscillates as a function of the applied flux. If the input current is more than I_c, then the SQUID always operates in the resistive mode. The voltage a function of the NERVE SIGNAL OR applied magnetic field and the period equal to ϕ_0.

a shunt resistance, R is connected across the junction to eliminate the hysteresis. In the case of copper oxide based <u>high-temperature</u> superconductors the junction's own intrinsic resistance is usually sufficient.

"So the project can work because we are using room temperature but we are looking at room temperature and intrinsic electrical resistance for a D.C. squid not a radio frequency squid with only one josephson junction."

"let us see if we can get the project to work for a r.f. squid at room temperature???"

Nuclear Physics, New Atomic Bomb, the Bionic Arm

RADIO FREQUENCY SQUID. m(t) nerve signal.

These frequency measurements can be easily taken, and thus the losses which appear as the voltage across the load resistor in the circuit are a 'periodic function of the applied magnetic flux', or nerve signal, with a period of ϕ_o, i.e. $\cos\{m(t) \pi/\phi_o\}$. For a precise mathematical description refer to the original paper by Erné et al.[5][10]

sin(a+b)-sin(a-b)=2cos(a)sin(b)

$$i = 2I_c \cos\left(\frac{\pi\Phi}{\Phi_o}\right) \sin\left(\varphi_1 + \frac{\pi\Phi}{\Phi_o}\right)$$

$$\Phi = \Phi_{ext} + \frac{LI_c}{2} \sin\left(\frac{\pi\Phi}{\Phi_o}\right) \cos\left(\varphi_1 + \frac{\pi\Phi}{\Phi_o}\right)$$

$= k\sin 2\pi o/o_o$ / /

m(t) is nerve signal implying

$k\sin(2\pi m(t)/n + LI_c/2\sin(2\pi m(t)/n)$
FORMAT.

The instantaneous frequency is

$2\pi d\{m(t)/n\}/dt +$ and $- LI_c/2$(max)

m(t) = nerve bundle signal.

An F.M. modulation electronics communication engineering electronic circuit, a passive electronic circuit,

 will give us this instantaneous frequency

$2\pi d\{m(t)/n\}/dt +$ and $- LI_c/2$(max)

.Note the following circuit diagram giving

dm(t)/dt as the output signal ; this circuit can go to 50°c. fig A. circuit diagram.

Nuclear Physics, New Atomic Bomb, the Bionic Arm

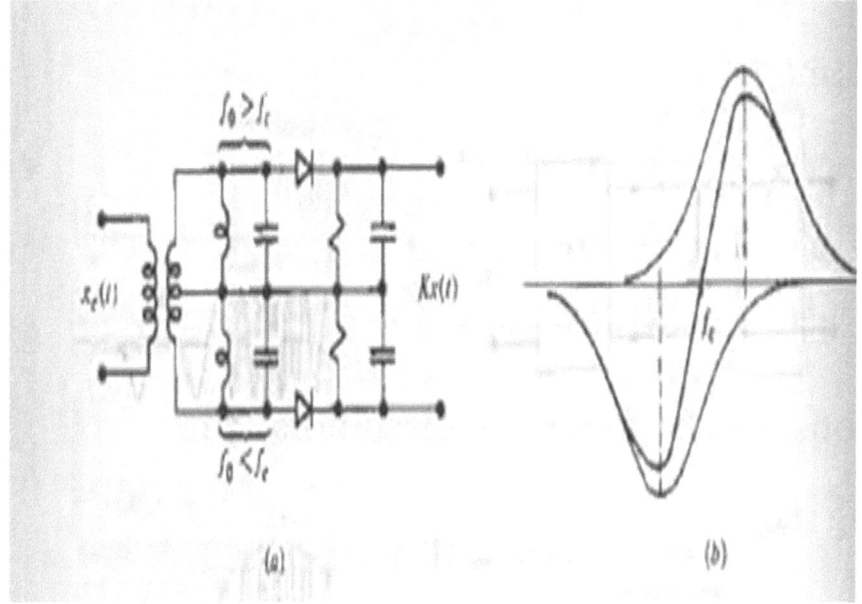

Nuclear Physics, New Atomic Bomb, the Bionic Arm

m(t) is the nerve bundle signal.

After fig A circuit diagram we have

d{m(t)}/dt.

Consider the following circuit diagram

Fig B.

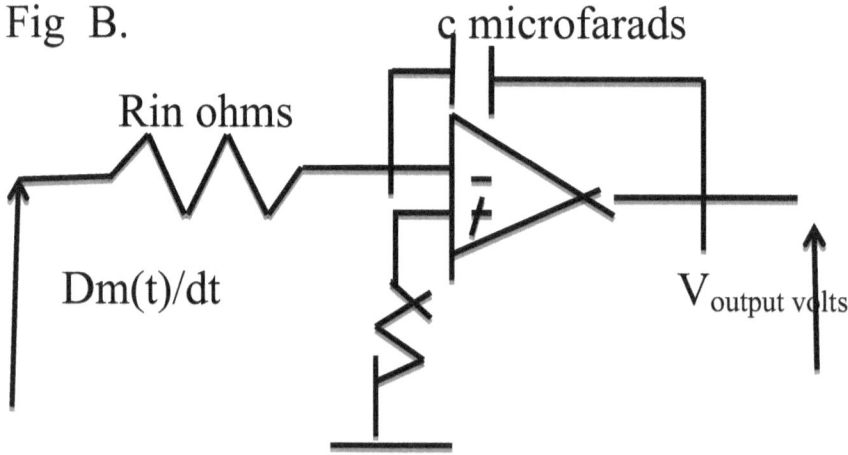

$V_{output\ volts}$ = m(t) the goal of the project, nerve signal

But there are problems to conquer.

The value of "c" and "rin" are chosen based on the fundamental FREQUENCY OR time period of the nerve pulse say 25 milliseconds according to boston general hospital.

But m(t) is a nerve bundle signal and we need to separate the nerve upper arm signal, forearm signal, hand signal, finger signal from the nerve bundle signal which also contains **"nerve sensory signals from the arm pit to the spine and brain."**

Mathematical cross-correlation from a healthy arm forearm movement gentle strength m(t) nerve signal to the disabled spine to shoulder nerve signal done by a microcomputer in the bionic arm electronics is proportional to the fourier transform of the two time nerve signals multiplied by each other.

m(t) nerve disabled bundle signal;

m(f) fourier of m(t).

m(t) forearm nerve signal from healthy arm =mf(t)

mf(f) fourier of mf(t)

Cross-correlation of m(t) with mf(t)

is "approximately" m(f) 'TIMESX' mf(f)

The micro computer of the bionic arm can work in the frequency domain.

Computer software or code in the microcomputer us assembler programmers prefer to call it compares m(f)mf(f) with $m(f)^2$

and tells the bionic arm if it is a $m(f)^2$ and moves the bionic arms forearm with d.c. motors.

m(f) is from the disabled arm.

mf(f) is from the healthy arm.

Nuclear Physics, New Atomic Bomb, the Bionic Arm

All healthy arm signals are in memory of the microcomputer and the healthy arm is not involved in the bionic arm.

The healthy arm is involved once in a calibration and memory storage of signals in the house of the disabled person and never again.

If a person has no arms we take the forearm attempted movement in both disabled arms and ensemble average the nerve bundle signals to produce a "calibrated forearm gentle movement" nerve bundle signal representing the nerve signal from any one of the spine to shoulders,

m(t)2dissabledarms = m(t)2dis

Nuclear Physics, New Atomic Bomb, the Bionic Arm

Strength in the bionic arm requires a different mathematical approach to fourier but at least the bionic arm now moves and picks an egg up here without smashing an egg or child.

A closed finger group of two or more fingers

Compress a spring in the bionic arm sending an electronic signal to the armpit to stimulate the nerve signal. What sensory muscle feedback is necessary for pouring a large pitcher of water out?

This study was prompted by recent evidence for the existence of positive force feedback in feline locomotor control. Our aim was to establish some basic properties of positive force feedback in relation to " load compensation", "stability", intrinsic muscle properties, and interaction with displacement feedback. In human subjects, muscles acting about the wrist and ankle were activated by feedback-controlled electrical stimulation. The feedback signals were obtained from sensors monitoring force and displacement. The signals were filtered to mimic transduction by mammalian tendon organ and muscle spindle receptors. We found that when muscles under positive force feedback were loaded inertially, they responded in a stable manner with increased active force. The activation attenuated the muscle stretch (yield) that would otherwise occur in the absence of feedback. With enough positive force feedback gain, yield could actually reverse. This behaviour, which we termed the affirming reaction, was reminiscent of the mammalian positive supporting reaction, a postural response elicited by contact of the foot with the ground. Muscles under positive force feedback remained stable, even when the loop gain (Gf) was set at levels of 2 or 3. In a linear system, if Gf > 1, instability occurs when the loop is closed. On further investigation, we found that Gf changed with joint angle: it declined as the load-bearing muscle actively shortened. We inferred that in closed-loop operation, the active muscles always shortened until Gf approached unity. In other words, the length-tension curve of active muscle ensures stability even when force-related excitation of motoneurons is very large. Concomitant negative displacement feedback reinforced and stabilized load compensation up to a certain gain, beyond which instability occurred.

The **Golgi tendon organ** (GTO) (also called Golgi **organ, tendon organ,** neurotendinous **organ** or neurotendinous **spindle**) is a proprioceptive sensory **receptor organ** that

Nuclear Physics, New Atomic Bomb, the Bionic Arm

senses changes in **muscle** tension. It lies at the origins and insertion of skeletal **muscle** fibers into the **tendons** of skeletal **muscle.**

frequency squid" or a "DC SQUID".

Electric field applied from a mosfet transistor, not a electronic signal, to nerve cell in a below research article. Moritz voelker and peter fromherz.

Electric field works for sensory, not magnetic field.

Key science

So a close proximity transistor device can provide sensory nerve feed back to the disabled person who would initially calibrate the feeling they get in eye movement monitoring of pouring out a pitcher of water. Originally the brain would get the feeling of lifting 20 pages of paper and taking of four pages at a time.

The brains memory would remember like musical memory the melody of feeling each time a pitcher of water is poured out.

Nuclear Physics, New Atomic Bomb, the Bionic Arm

It is the auhors opinion that artificial intelligent sensory feedback is remembered in the memory of sound area of the brain. 17,500 nerves go to this area of t he brain.

The bionic arm would have a transmitter near the shoulder area with a high antenna gain antenna pointing to the spine. Lets examine the frequency of the nerve motor <u>nerves and sensory nerves.</u>

Rfsquid temperature extension range

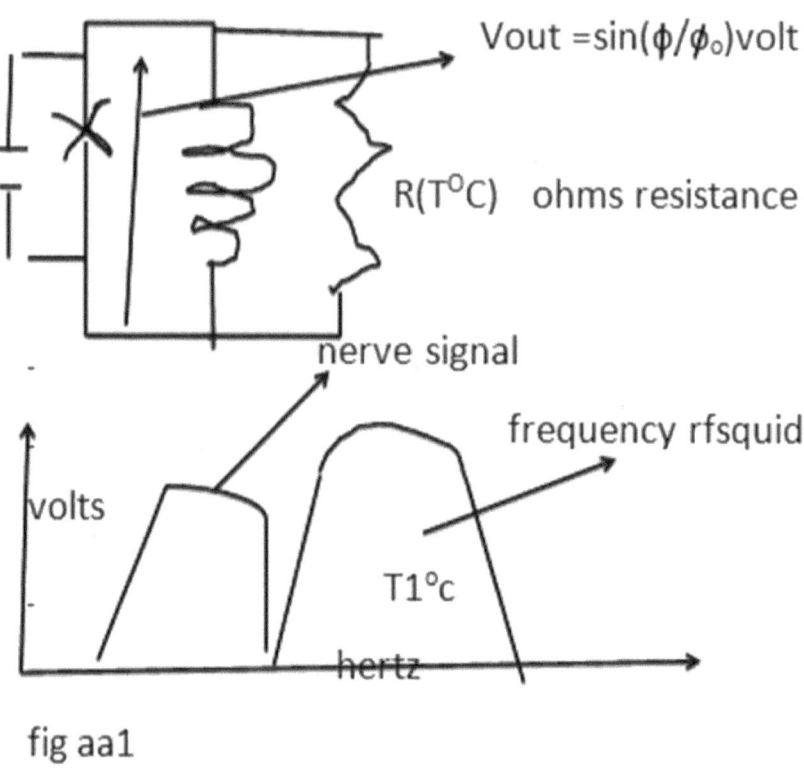

fig aa1

the nerve signal frequency is lower than the rfsquid and does not get to the output electrically due to high temperature T1°C

Nuclear Physics, New Atomic Bomb, the Bionic Arm

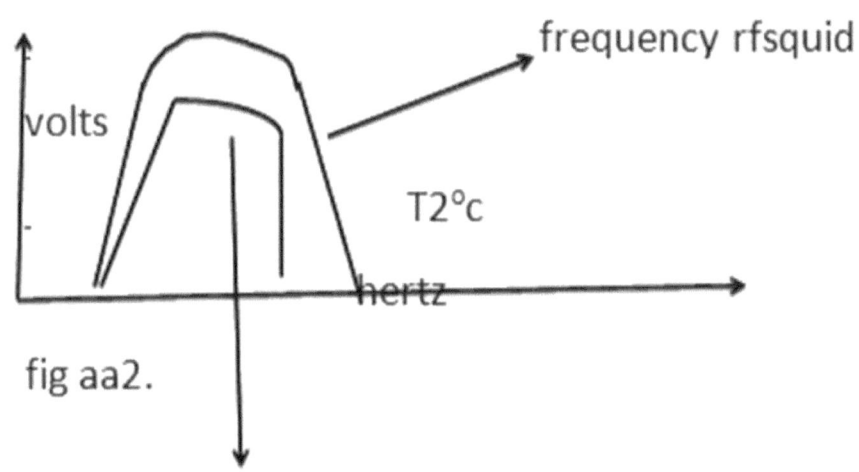

fig aa2.

nerve signal at lower temperature T2°C

the nerve signal goes to the electrical output and is picked up by the electronics.

The capacitor c, inductor L and resistor $R(T^oC)$ are the material properties of the r.f.squid.

In fig aa1 , fig aa2.

Nuclear Physics, New Atomic Bomb, the Bionic Arm

Take the new invention of the igla squid as follows.

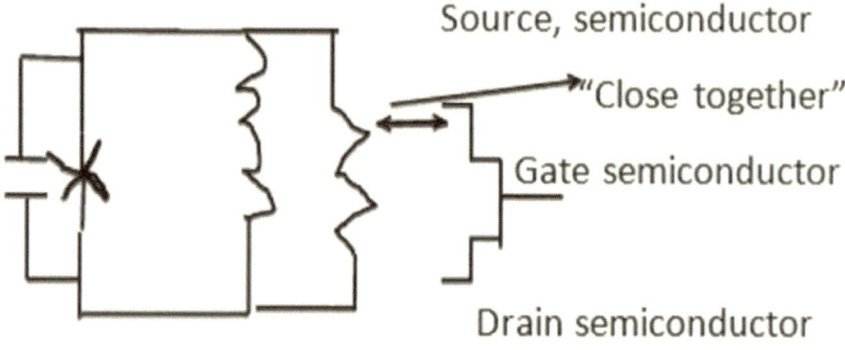

The mosfet transistor next to the rf squid places an "electric field" along the "path" of the rfsquid. Reason Maxwell see equations.

The physics here needs to show improved temperature characteristics. Refer to Andrew Igla's physics book Maxwell equations.

MAXWELLs EQUATIONS to follow.

1. $\nabla \cdot \mathbf{D} = \rho_v$
2. $\nabla \cdot \mathbf{B} = 0$
3. $\nabla \times \mathbf{E} = -\dfrac{\partial \mathbf{B}}{\partial t}$
4. $\nabla \times \mathbf{H} = \dfrac{\partial \mathbf{D}}{\partial t} + \mathbf{J}$

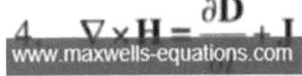

The theory of whole superconductivity predicts that when a metal goes superconducting negative charge is expelled from its interior towards the surface. As a consequence the superconductor in its ground state is predicted to have a nonhomogeneous charge distribution and an "outward" pointing electric field in its interior.

$$\frac{\partial}{\partial t}j = \frac{e^2 n}{m}E$$

units coulombs per second squared

CALTECH SUPERCONDUCTIVITY ARTICLE

$$\nabla \times E + \frac{1}{c}\frac{\partial B}{\partial t} = 0$$

$\nabla \times E = -B'(t)$

$\nabla \times B/u = J$

$\nabla \times B'(t)/u = j'(t) = e^2 \, n/m \, E$

$j'(t) = e \, n/m \, E$

$\nabla \times \nabla \times E = 1/c \, j'(t) = \{e^2 \, n/m\} \, E$

$\nabla^2 E = \{1/c\} \, j'(t) = \{e^2 \, n/m\} \, E$

E can mathematically be worked out here simply in a superconductor.

Let us look at recombination of electrons in a superconductor.

The E field from a mosfet transistor is at an angle to the Rf squid and a small E field in intensity of frequency 'w'

$$E = E_O \sin(wt - bx)$$

and is in all directions in the superconductor.

The electric field already proven creates electric charge along the surface of the superconductor given by formulae

$\int j'(t)dt$. **This electric charge is 'n' in the above formulae charge per volume.**

But this is not seen electronically at the R.F. squid with frequency 'w'. The r.f. squid is a

tuned electronic circuit that filters out the signal with frequency 'w'.

So Frequency 'w' of the electric field is outside the frequency band of the r.f. squid.

But at higher temperatures of the r.f.squid has electric charge along the surface of the superconductor "statistically" available with the use of the electric field of the mosfet transistor close to the rf squid. This allows an extension of temperature range over a standard r.f. squid. This is the igla squid.

Nuclear Physics, New Atomic Bomb, the Bionic Arm

In a super conductor the electric field point radially outward. Statistically along the surface of a superconductor the time varying electric field exists statistically with a mosfet transistor placed close to the surface of the rf squid at an angle. Why at an angle?

Maxwell equation $\nabla \times E$ is evaluated in a spatial loop around the surface of the R.F. squid. The electric field needs to exist in all unit vector axis along the surface of the r.f. squid loop for sensitivity in physics. Hence the need for an angle of the mosfet transistor to the squid.

Temperature range of the rf squid increased due to charge availability for physics tunnelling across the energy barrier effect.

The electric field from the mosfet transistor does not drive the charge across the insulating barrier of the rf squid because the frequency 'w' is out of the rf squid range.

The electric surface charge from the electric field from the mosfet transistor simply provides statistically electric charge at a higher temperature range of operation of the rf squid.

How too separate nerve signals in nerve bundle group?

The r.f. squid collects the combined magnetic fields of the nerve bundle. Previous nerve separation signals were about correlation in this project so **far in time.**

But fourier transforms mathematically of the combined nerve signal allows frequency graphs causing a unique relation between volts from the nerve signal and frequency.

This is the correct logical path for nerve bundle group separation. Allowing for smooth movement of the bionic arm robotically. Using mathematical CORRELATION for nerve bundle separation as previously done will cause non smooth movement of the bionic arm but with calibration as already mentioned will still work in this project.

The mathematics after the fourier is the derivative of fourier and integral of fourier all giving information of nerve bundle group separation. Extracting the unique numbers for each nerve bundle group signal in the end.

Nuclear Physics, New Atomic Bomb, the Bionic Arm

An analogy to this is the bionic ear.

A group of hairs is connected to group of nerves allowing a unique number to be interpreted by the brain in the 17,600 nerves to the brains lower area memory of sound region. And so beethoven a deaf person wrote brilliant music.

The area under the fourier graph is the non stochastic time varying variance and average volt nerve signal squared. Parsevals theorem mathematically 1800.

The differential in fourier is the equivalent of a hair connected to many nerves and a nerve connected too many hairs in the bionic ear.

Nuclear Physics, New Atomic Bomb, the Bionic Arm

The differential is the fourier graph multiplied by frequency giving a first order fourier.

The graph fourier is multiplied by frequency f, frequency f squared f^2, f cubed f^3, to f^7.

All giving the bionic arm in its micro computer memory of movement like the bionic ear memory of sound with each hair connected to seven nerves and a nerve connected to memory of sound, beethovens symphany.

An intelligent bionic arm.

Software using Fourier analysis in medicine and radar allow the fourier of a nerve signal to happen in the microcomputer of the bionic arm.

The derivatives mathematically of the nerve signal is done via the microcomputer of the bionic arm. This involves multiplying the fourier frequency graph digitally by frequency.

Then the graph of the fourier frequency system is multiplied by frequency again to achieve a higher order derivative of the original nerve signal. This repeated seven times.

Nuclear Physics, New Atomic Bomb, the Bionic Arm

Then a signature formulae algebrically is developed to create numbers to logically tell the microcomputer on the bionic arm to move uniquely and dynamically and through a set of signature numbers guide the bionic arm to cope and smoothly pour the pitcher of water out without jerking movement.

The signature numbers are:

Fourier is $F\{m(t)\} = a11$ at frequency $f11$ where $m(t)$ is a nerve signal.

Fourier first order derivative is f times $F\{m(t)\} = a21$ at frequency $f21$ where $m(t)$ is a nerve signal and f is frequency.

And so on until ...

Fourier seventh, 7th, order derivative is f too the power of seven, 7, times $F\{m(t)\} = a71$ at frequency f71 where m(t) is a nerve signal and f is frequency.

But all the above examples are at frequency 1 in their respective graphs of particular order derivative, starting with a fourier of the m(t) signal or nerve signal

. i.e. f11,f21,…,f71.

Meaning frequency 1, first order derivative, second order derivative, …., seventh order derivative.

Now f21, f22,… ,f72 are frequency 2 on each fourier graph representing fourier of m(t) and first order derivative and so on too seventh order derivative.

There will be sixty four frequencies per graph, each graph representing normal fourier of m(t) nerve signal and fourier of first order derivative of m(t) nerve signal and up to seventh order derivative of m(t).

The group of numbers allow the microcomputer in the bionic arm to move smoothly in a certain way for a task.

During a bionic arm task the bionic arm updates its current time task to adjust to the environment.

Say a pitcher of water suddenly looses most of its water by a other person bumping the person with the bionic arm or the person with the bionic arm is wrestling.

Nuclear Physics, New Atomic Bomb, the Bionic Arm

There will be a maximum dynamic movement to load weight change allowable in the microcomputer of the bionic arm.

Just like a dynamic code changing situation in intelligence when digital pursuit is active.

STRENGTH MAXIMUM of the bionic arm is applied swiftly or slowly.

m(t) is the nerve signal from the spine to the arm pit picked up by weak magnetic fields that tell the arm to grasp a bowling ball.

Nuclear Physics, New Atomic Bomb, the Bionic Arm

The frequency spectrum of m(t) ; volts versus frequency ; can be intense in the upper frequency arena for a quick grasp or smaller for a slow grasp.

The first order linear differential of m(t) and second order linear differential of m(t) nerve signal has a more intense graph at the higher frequencies of the frequency spectrum of $dm(t)/dt$, $d^2m(t)/dt^2$ for strength of lift.

Fourier of $dm(t)/dt$, $d^2m(t)/dt^2$ is

$F\{ dm(t)/dt \} = m'(f)$

$F\{ d^2m(t)/dt^2 \} = m''(f)$

Graph of strong mechanical grasp swift or slow enacted.

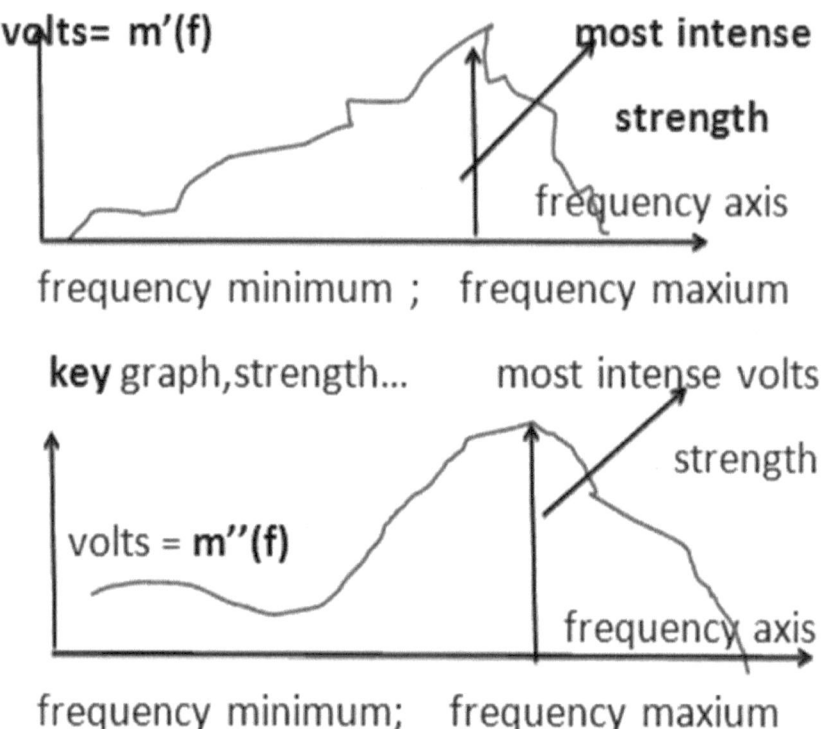

Strong mechanical grasp explain.

A strong mechanical grasp requires the higher frequencies of the first and second order derivative of the nerve signal m(t) and actual nerve signal m(t) to have most of the area of the first and second order derivative graph of m(t) and most o f the area of the actual graph m(t).

The memory of the bionic arm's micro computer will get the nerve signal and process a strong mechanical grasp.

Usually a fast fourier transform is done on the microcomputer of the nerve signal.

Nuclear Physics, New Atomic Bomb, the Bionic Arm

Strong mechanical grasp changing in time.

Pouring a heavy pitcher of water out requires a strong mechanical grasp with varying strength.

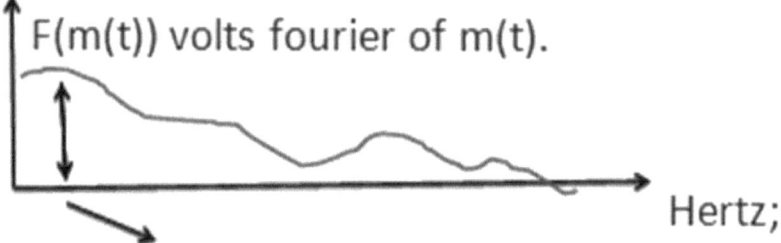

Consider the fourier of the nerve signal m(t), F(m(t)) volts fourier of m(t). Hertz; Lower frequencies involved Frequency.

Force $= mdv/dt = md^2x/dt^2 = kx$; x being height of water in jug pitcher.

Lower frequencies are involved here.

Let $x = \sin bt$; $b^2 = k/m$; $b = \sqrt{k/m}$; $x = a \sin\sqrt{k/m}\,t$; frequency $= \sqrt{k/m}$ lower frequency.

Nuclear Physics, New Atomic Bomb, the Bionic Arm

So pouring a pitcher of water out requires a strong mechanical grasp set of frequencies and a lower frequency in the nerve signal m(t). This is a mathematical linear set of frequencies.

Nuclear Physics, New Atomic Bomb, the Bionic Arm

The design of the bionic arm will be similar to a biological arm with bends, observe here.

The arm... fig 23...

Upperarm forearm Hand finger, thumb

Spring moving devices move components relative to each other.

No magnetic field here IN sensory data.

Important posterior nerve electric field via the axillary nerve at the armpit fact and figure.

Electric field created under the armpit by a fet transistor taped to the arm pit electrically

plugged into the bionic arm will increase in signal strength as the "mechanical weight" of the "mechanical grasp" increases. The armpit Electric field source creator will be in a rubber chassis body.

Biomechanics of the arm will have dc motors and springs to tension one part of the arm to the other too the hand to the finger. This will be a battery operated bionic arm.

So ends the project of the modern bionic arm Andrew Maxim Igla.

Nuclear Physics, New Atomic Bomb, the Bionic Arm

Bibliography

Nuclear Reactions

https://2012books.lardbucket.org/books/principles-of-general-chemistry-v1.0/s24-02-nuclear-reactions.html

Alpha Decay

http://nuclearpowertraining.tpub.com/h1019v1/css/Beta-Decay-48.htm

Beta Decay

http://nuclearpowertraining.tpub.com/h1019v1/css/Beta-Decay-48.htm

Neutron Moderator

https://en.wikipedia.org/wiki/Neutron_moderator

Neutron Moderator

http://www.nuclear-power.net/neutron-moderator/

Nuclear Fission

http://hyperphysics.phy-astr.gsu.edu/hbase/NucEne/fission.html

Average Number of Neutrons Liberated in Fission

http://nuclearpowertraining.tpub.com/h1019v2/css/Table-1-Average-Number-Of-Neutrons-Liberated-In-Fission-31.htm

Axillary nerve

http://www.statemaster.com/encyclopedia/Axillary-nerve

Positive Force Feedback Control of Muscles

https://pdfs.semanticscholar.org/dbbf/c9b5d6fa94963d48527293221630l3293b12.pdf

Brachial Plexus

https://en.wikipedia.org/wiki/Brachial_plexus

Nuclear Structure

http://www2.lbl.gov/abc/Basic.html

www.ingramcontent.com/pod-product-compliance
Lightning Source LLC
Chambersburg PA
CBHW030847180526
45163CB00004B/1488